YOUR KNOWLEDGE HAS VALUE

- We will publish your bachelor's and master's thesis, essays and papers

- Your own eBook and book - sold worldwide in all relevant shops

- Earn money with each sale

Upload your text at www.GRIN.com
and publish for free

Bibliographic information published by the German National Library:

The German National Library lists this publication in the National Bibliography; detailed bibliographic data are available on the Internet at http://dnb.dnb.de .

This book is copyright material and must not be copied, reproduced, transferred, distributed, leased, licensed or publicly performed or used in any way except as specifically permitted in writing by the publishers, as allowed under the terms and conditions under which it was purchased or as strictly permitted by applicable copyright law. Any unauthorized distribution or use of this text may be a direct infringement of the author s and publisher s rights and those responsible may be liable in law accordingly.

Imprint:

Copyright © 2017 GRIN Verlag, Open Publishing GmbH
Print and binding: Books on Demand GmbH, Norderstedt Germany
ISBN: 9783668521568

This book at GRIN:

http://www.grin.com/en/e-book/373835/seismic-performance-of-rc-frame-structures-with-semi-interlocked-masonry

Mangeshkumar Shendkar, M. L. Waikar

Seismic Performance of RC Frame Structures with Semi-Interlocked Masonry and Unreinforced Masonry Infill

A Comparative Study

GRIN Publishing

GRIN - Your knowledge has value

Since its foundation in 1998, GRIN has specialized in publishing academic texts by students, college teachers and other academics as e-book and printed book. The website www.grin.com is an ideal platform for presenting term papers, final papers, scientific essays, dissertations and specialist books.

Visit us on the internet:

http://www.grin.com/

http://www.facebook.com/grincom

http://www.twitter.com/grin_com

Comparative study on RC frame structures with SIM infill and URM infill

Mangeshkumar Shendkar[1], M.L.Waikar[2]

[1] P.G. Student, Department of Civil Engineering, SGGSIE&T, Nanded, India
[2] Professor, Department of Civil Engineering, SGGSIE&T, Nanded, India

Abstract: The main objective of the study was to investigate the importance of inter-locked brick infill in multi-storied buildings it uses masonry panels made of dry stack interlocking masonry units capable of relative sliding in plane and interlocked to prevent sliding out-of-plane. The major objective of the system is to improve the earthquake performance of framed structures with masonry panels acting as energy dissipation devices (EDD). An analytical program conducted to evaluate the behavior of different framed masonry panels. It was found that semi-interlocked masonry (SIM) panels have significant energy dissipation capacity due to friction between the masonry units. This paper presents the results of a numerical simulation of earthquake vibrations on a multi-storey reinforced concrete frame with different infills namely SIM & URM panel.

Key words: Semi-interlocked masonry, nonlinear static analysis, Energy dissipation device etc.

INTRODUCTION

Recently, the most common structural system for both residential and office buildings consists of multi-level framed structures are masonry infilled RC frames so it is so important to determine the earthquake behavior of RC structures with infill walls under seismic load. Nonlinear structural analyses and finite element method are used to determine the earthquake behavior of structures with infill walls. Masonry is one of the most popular construction materials. It has many excellent properties and proven durability. Over time, masonry has evolved from a material used for massive structural walls, which work mainly in compression, to more slender walls, which can also be subjected to considerable tension and shear. However, modern slender unreinforced masonry (URM) walls have poor earthquake resistance due to their high mass and with low tensile and shear strength. The design of masonry with improved earthquake resistance presents a challenge for structural engineers .SIM is a conceptually new type of framed masonry built of dry stack interlocking units.SIM units are capable of relative sliding in plane and locked against relative movement out-of-plane. Interlocking brick system in building works is a fast and cost effective construction system and provides good solution in construction. Hence, there is a need to hasten the effort to determine the effectiveness of using interlocking brick in construction system.

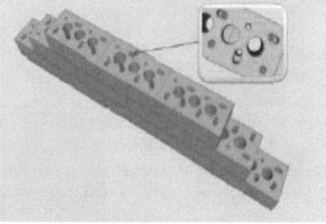

a) Topological   b) Mechanical
Fig 1.1 Different methods of semi-interlocking

MATERIALS AND METHODS

1 Introduction:

Nonlinear static analyses of interlocked brick and standard brick infilled frames and bare frame specimen were performed under finite element software Seismostruct. The prominent features of this software are capability of predicting the large displacement behavior of space frames under static or dynamic loading, taking into account both geometric nonlinearities and material inelasticity. The monotonic lateral loads applied to the top of the columns, modeled as in software. Materials, sections, elements, restraints and loading type are the characteris that affect specimen behavior substantially. Models (material models, element models) which related with th characteristics have numerous empirical and physical parameters.

2.2 Problem statement:

For this study, a 5-storey with 5 bays two dimensional frame (Each bay span 4 m) and floor height 3.0m, regula plan is considered. This building is considered to be situated in seismic zone 'v' and designed in compliance to Indian Code of Practice for Earthquake Resistant Design of Structures. The building is modeled using softv seismostruct. Model is studied for comparing the performance of RC frame structures with semi-interloc masonry panel and unreinforced masonry panel for different models as follow:
1) Bare frame
2) URM (open ground RC frame)
3) URM (only side bay infilled at ground of RC frame)
4) URM (Full infilled RC frame)
5) SIM (Full infilled RC frame)
6) SIM (open ground RC frame)
7) SIM (only side bay infilled at ground of RC frame)

Table 2.1 Structural details of RC frame

Type of structure	Ordinary moment resisting frame
Seismic zone	V
Number of stories	Five, G+4
Floor Height	3m
Bay length	4m
Infill wall	URM wall-113mm SIM wall- 110mm
Type of soil	Soft soil
Size of column	450mmX450mm ,450mm X 600mm
Size of beam	300x500mm ,300x450 mm ,300x 300mm
Depth of slab	100mm
Live load	2.5 kN/m^2
Material	M30 grade & Fe 500 reinforcement
Damping in structure	5%
Importance factor	1.5

2.3 Analytical study

2.3.1 Introduction

Nonlinear static analyses of interlocked brick and standard brick infilled frames and bare frame specir performed under finite element software Seismostruct. The prominent features of this software are capabili predicting the large displacement behavior of space frames under static or dynamic loading, taking into acc both geometric nonlinearities and material inelasticity

2.3.2 Material models

A) Concrete Model (Mander et. al – concrete model):
This model is a uni-axial nonlinear constant confinement model that the effects of confinement provided b lateral transverse reinforcement

B) Steel Model (Menegotto –pinto steel model)
This is a uni-axial steel model initially programmed by Yassin [1994]. Its employment should be confined t modeling of reinforced concrete structures, particularly those subjected to complex loading histories. Stress-s

relationship proposed by Menegotto and Pinto [1973], coupled with the isotropic hardening rules proposed by Filippou et al. [1983]

2.3.3 Element classification for column, beam and infill

A) Inelastic Force-Based Frame Element type (infrmFB) for column and beam:
This is the force-based 3D beam-column element type capable of modeling members of space frames with geometric and material nonlinearities. As described in the Material inelasticity paragraph, the sectional stress-strain state of beam-column elements is obtained through the integration of the nonlinear uni-axial material response of the individual fibers in which the section has been subdivided, fully accounting for the spread of inelasticity along the member length and across the section depth. Element infrmFB is the most accurate among the four inelastic frame element types of SeismoStruct, since it is capable of capturing the inelastic behavior along the entire length of a structural member, even when employing a single element per member. Hence, its use allows for very high accuracy in the analytical results

B) Inelastic Infill Panel Element (infill):

Each infill panel element represented by four axial struts and two shear springs, as shown in Figure .This element is able to define with three groups of parameters. First group is about physical characteristics of infill panel, second group is about compression/tension struts defined by strut curve parameters, and third group is about shear spring that defined by shear curve parameters.

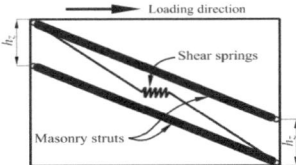

Fig 2.1 Inelastic infill panel element

2.3.4 Data Compilation and Calculation

Lumped mass is calculated and applied for each node which is due to the dead weight of the floor slab and the infill walls. Reinforcement in beam and column sections for the structure are calculated according to analytical results of frame from SAP-2000 using gravity load & seismic load condition with M30 concrete and Fe500 steel reinforcement. These sections are assigned to the simulation of the structure made in Seismostruct and lumped masses are also assigned to each node. Thus the structure is simulated in Seismostruct with 5stories-5 bays two dimensional frames with different infill conditions. This structure is loaded from x-axis to get the performance curves in respective axis.

Table 2.2 Seismic weight calculation

FLOORS	SEISMIC WEIGHT
Ground floor	656.42 kN
First floor	656.42 kN
Second floor	614.68 kN
Third floor	614.68 kN
Fourth floor	434.60 kN

On the basis of this seismic weight calculation,
Design Base shear = $A_h \times W$ = 0.225x 2976
= 669.8 KN

This base shear is shared amongst each floor as:
Loading along x-axis:
1. 14.88 kN (Slab Level 1)
2. 59.54 kN (Slab Level 2)
3. 125.46 kN (Slab Level 3)
4. 223.05 kN (Slab Level 4)
5. 246.41 kN (Slab Level 5)

3. RESULTS AND DISCUSSIONS

A) Pushover Curves of different RC in-filled frames

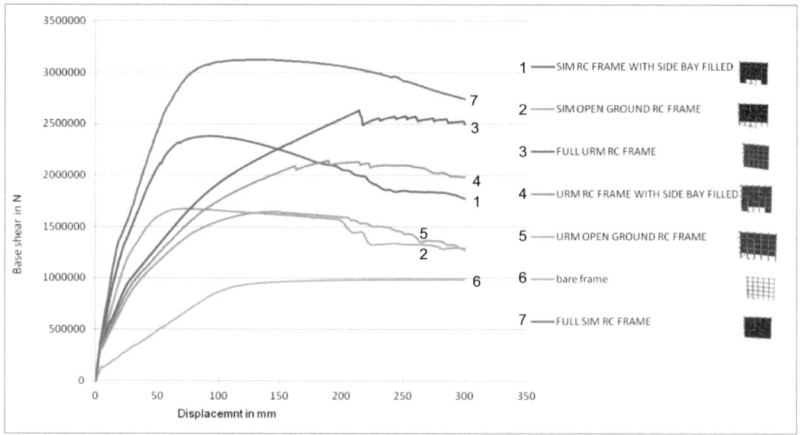

Graph 3.1 Comparison of pushover curves for Different frames

B) Base shear:

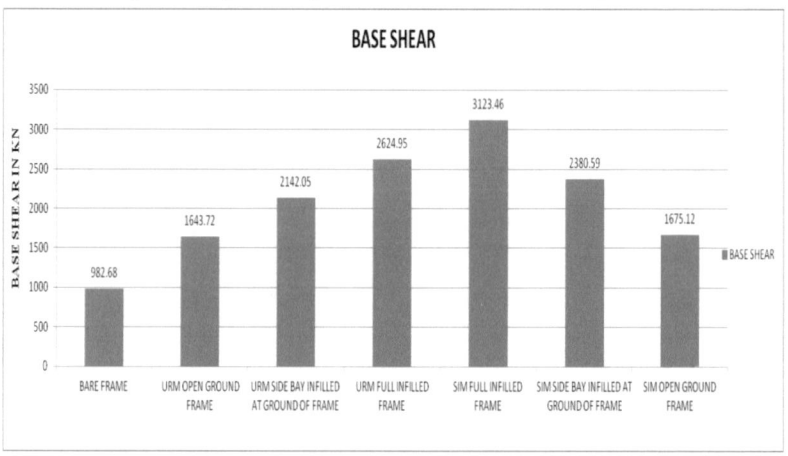

Graph 3.2 Comparison of Base Shear for Different frame

C) DUCTILITY:

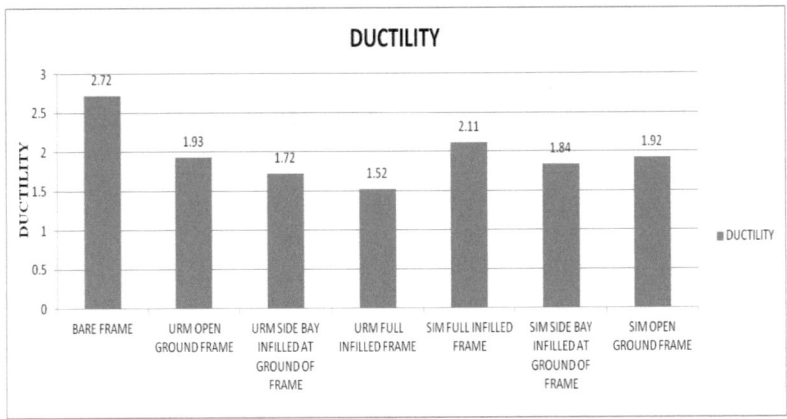

Graph 3.3 Comparison of Ductility for Different frames

4. CONCLUSIONS
1. Base shear is maximum in SIM infilled frame as compared to URM infilled frame.
2. Ductility is maximum in SIM infilled frame as compared to URM infilled frame.
3. Ductility is higher in bare frame as compared to all other frames.
4. Ductility reduces when frame infilled with SIM and URM panel.
5. SIM panel acts as a energy dissipation device in structure because it dissipates more energy through relative shear sliding mechanism
6. SIM infilled frame gives fast & economic construction as compared to URM infilled frame

REFERENCES

Crisafulli FJ. (1997) "*Seismic behaviour of reinforced concrete structures with masonry infills*". Ph.D. Thesis, University of Canterbury, New Zealand

E. Smyrou , C. Blandon , S. Antoniou , R. Pinho , F. Crisafulli [2011] "*Implementation and verification of a masonry panel model for nonlinear dynamic analysis of in-filled RC frames*"

Francisco J. Crisafulli and Athol J. Carr [2007] "*Proposed Macro-Model For The Analysis Of Infilled Frame Structures*" The New Zealand Society For Earthquake Engineering, Vol. 40.

Ibrahim Serkan Misir [2014] "*Potential Use of Locked Brick Infill Walls to Decrease Soft-Story Formation in Frame Buildings*" American Society of Civil Engineers

Koray Kadas [2006] "*Influence of Idealized pushover curves on seismic response*"

Lini M Thomas, Kavitha P.E. [2016] "*Effect of Locked Brick Infill Walls on the Seismic Performance of Multistoried Building*" IOSR Journal of Mechanical and Civil Engineering (IOSR-JMCE)

Tarek M. Alguhane, Ayman H. Khalil, M.N.Fayed, Ayman M. Ismail[2015)]"*Seismic assessment of old existing RC buildings with masonry infill in madinah as per ASCE*" International Journal Of Civil, Environmental, Structural, Construction And Architectural Engineering Vol:9, No:1

Zhiyu Wang,Yuri Totoev,Adrian Page,Willy Sher and Kun Lin [2015] "*Numerical simulation of earthquake response of multi-storey steel frame with SIM infill panels*" Advances in structural engineering and mechanics (ASEM)

Z. Wang , Y. Totoev & k.Lin [2016] "*Non-linear static analysis of multi-storey steel frame with semi-interlocking masonry infill panels*" Brick and Block Masonry – Trends, Innovations and Challenges – Modena, da Porto & Valluzzi (Eds)© 2016 Taylor & Francis Group, London, ISBN 978-1-138-02999-6

YOUR KNOWLEDGE HAS VALUE

- We will publish your bachelor's and master's thesis, essays and papers

- Your own eBook and book -
  sold worldwide in all relevant shops

- Earn money with each sale

Upload your text at www.GRIN.com
and publish for free